Practice Tests for Math GOAL 2 Level C, Forms 925M and 926M

Helping Learners Develop Mathematical Thinking Skills, Approach Math with Confidence, and Sharpen their Test-Taking Ability

By

TABLE OF CONTENTS

PREFACE

Dear Instructors,

This book entails four (4) practice tests and is designed to prepare adult learners for the CASAS Math GOALS 2, Level C Forms 925M and 926M. The practice tests align with the CASAS Competencies and meet the rigorous requirements of the College and Career Readiness Standards (CCRS), the National Reporting System (NRS), and the Workforce Innovation and Opportunity Act (WIOA).

Adhering to the CASAS Math GOALS 2 test blueprint, the practice tests assess learners' understanding of the following math areas: *Number Sense and Operations, Consumer Economics, Algebraic Thinking, Geometry, Data Analysis and Statistics, and Pure Mathematics.*

More importantly, this resource increases learners' confidence, guides them to reflect on their learning and progress, and helps them transfer their knowledge to other contexts. Each practice test includes real-world activities that promote deep understanding and practical application of mathematical concepts. An answer key also accompanies each test.

By using this resource, you can save time, distribute practice sessions over several weeks, and assess and reinforce your learners' understanding of math functions and concepts. To order class sets, visit cbledu.com.

The CBL Team,

Your Partner in Student Learning

INTRODUCTION

Dear Math Students,

This resource will help you develop and reinforce your math skills and test-taking ability. It will prepare you for the CASAS Math GOALS 2 Level B test. The practice tests will assess your understanding of the following math areas: *Number Sense and Operations, Consumer Economics, Algebraic Thinking, Geometry, Data Analysis and Statistics, and Pure Mathematics.*

Important Strategies:

Follow the strategies below to develop and reinforce your mathematical thinking skills and test-taking ability.

1. Study and master the four operations (addition, subtraction, multiplication, and division). Learn several strategies to compute and perform operations.

2. Study and master the multiplication table by reviewing it at least once daily (5 to 10 minutes).

3. Look up the meanings of math concepts (e.g., *sum, product, quotient, fraction, triangle*). Try to describe their meanings in your own words.

4. Seek to understand math ideas or the big picture before practicing the details or simple exercises. You can do that by using YouTube videos or Khan Academy.

5. Connect math ideas and concepts to real-world objects or situations. Ask your instructors for real-life examples.

6. Ask clarifying questions to ensure you understand everything before your class ends.

7. Practice solving word problems weekly (20-30 minutes) without distraction (TV, PC, cellphone, noise).

8. Solve math operations and problems on paper. Always show your work—including your strategies or reasoning—on paper.

9. Study and practice math in a group or with a classmate. Discuss your math solutions and strategies.

10. Explain math ideas and concepts to yourself or someone else orally. After doing this orally, you can also do it using drawings and writing.

11. Always reflect on your progress and strategies. After each practice session or test, identify what works well and why you make certain mistakes. Review and focus on practicing math ideas and concepts you don't understand well.

12. Celebrate your achievements. Any increase in math knowledge is an achievement.

Remember, math skills are essential for success in various aspects of your life, including community involvement, managing family finances, and professional advancement. By committing to completing the practice tests in this book, you'll be setting yourself up for success in your academic pursuits and beyond.

Let's get to work!

HOW TO APPROACH MATH WITH CONFIDENCE

Here are nine (9) practical ways you can overcome math fear and anxiety and build confidence while using this math book:

1. **Start Small:** Begin with easier problems that you can solve to build your confidence before solving harder ones.

2. **Use the book's Strategies:** Take advantage of this resource's strategies and practice tests. They are designed to help you understand, practice, and sharpen your math skills.

3. **Set Small Goals:** Break your math studies into small, achievable goals. Celebrate when you reach these goals to motivate yourself.

4. **Practice Regularly:** Consistent practice makes learning math more manageable. Try to work on math problems a few times a week.

5. **Take Breaks:** If you feel overwhelmed, take a short break. Come back to the math problems with a clear mind.

6. **Ask for Help:** Don't hesitate to seek help when you need it. Ask a teacher or a classmate, or use online resources if you're stuck.

7. **Stay Positive:** Keep a positive attitude about math. Remind yourself that you can handle it and that it's okay to make mistakes as you learn.

8. **Understand, Don't Memorize:** Focus on understanding the math ideas and concepts rather than just memorizing formulas. This understanding will make you feel more confident in solving math problems and taking math tests.

9. **Visualize Success:** Picture yourself successfully solving problems and understanding concepts. This visualization can boost your confidence.

By following these strategies, you will be able to study well and practice math with more confidence.

PRACTICE TEST #1

You have 70 minutes to answer 36 questions.

1. The selling price of an item is 1.34 times the price the store paid. If the selling price is $335, what did the store pay?

   A. $448.90

   B. $250

   C. $248.55

   D. $290.99

2. The cost of a bike is $315.85. What is the cost of a dozen bikes?

   A. $3,790.20

   B. $3,590.95

   C. $3,828.50

   D. $3,330.08

3. Susan's car gets 36.5 miles per gallon on the highway. If her fuel tank holds 40 gallons, how far can she travel on one full tank of gas?

   A. 1,500 miles

   B. 1,440 miles

   C. 1,460 miles

   D. $1,536 miles

4. Which of the following is equivalent to 1.25?

   A. 5/4

   B. 3/2

   C. 1 + 1/2

   D. 1 x 0.25

5. Corey bought a used car for $8,000. He paid a $2,000 down payment and will pay the balance in 12 equal payments over the year. How much will he pay for each month?

   A. $600

   B. $550

   C. $725

   D. $500

6. Compute 0.75 x (1, 200 – 800)?

   A. 300

   B. 350

   C. 400

   D. 250

7. An ice cream machine makes 480 quarts of ice cream in 8 hours. How many quarts could be made in 10 hours?

   A. 520 qt.   C. 600 qt.

   B. 580 qt.   D. 640 qt.

8. What is the value of M in the following proportion?

$$\frac{M}{3} = \frac{11}{6}$$

   A. 6   C. 5.5

   B. 8   D. 6.2

> James's school fee cost $7,620 for this term, and financial aid covered 75% of that amount.

9. How much did financial aid cover?

   A. $5,700   C. $5,300

   B. $6,125   D. $5,715

10. How much does James still have to pay?

   A. $1,895   C. $1,860

   B. $1,905   D. $1,920

11. Which of the following is true?

   A. 100% of 50 = 50   C. 1% of 1 = 0.1

   B. 50% of 50 = 5   D. 20% of 20 = 5

12. Simplify the following expression:

$$\left(\frac{2^4 \cdot 2^8}{2^9 \cdot 2^2}\right)^{-2}$$

   A. 1/2   C. 2^{-1}

   B. 2^3 D. 1/4

13. What is k?

$$a^6 \cdot a^k = a^{18}$$

A. -12 C. 12

B. 8   D. 3

14. What is the value of x?

$$50\% \text{ of } x = 17$$

A. 8.5   C. 34

B. 32   D. 48

Look at the following receipt:

STAPLE-STORE

LOW PRICES, EVERY DAY
2344, Staple Furniture Road
Furniture City, CA, 211232

SALE		27981349442287755000
		10/17/2020 17:23

QTY	SKU	PRICE
1	HAND TOWEL	
	023404213519	2.97 N
4	Office Chair	
	069005841315	359.56 N
1	Office Table	
	030424458834	120.89 N
SUBTOTAL		$483.42
	Simple Tax 10.25%	$50.27
TOTAL		?

CREDIT
Card No .: xxxxxxxxxxxx 9999
Chip Read
Auth No .: 688880
AID .: 6166TF6V3RQ4

15. How many items were purchased?

A. 3   C. 4

B. 5   D. 6

16. What is the total amount?

A. $533.69   C. $523.79

B. $483.42   D. $512.69

17. What is the cost of a dozen office chairs?

 A. $359.56

 B. $915.05

 C. $1,078.68

 D. $1,438.24

18. What is the cost of ten hand towels?

 A. $18.97

 B. $29.70

 C. $25.64

 D. $31.25

19. Frank walks 5 miles in 4/5 of an hour. What is the unit rate in miles per hour?

 A. 7.55 miles per hour

 B. 6.72 miles per hour

 C. 7.50 miles per hour

 D. 6.25 miles per hour

20. Ashley can mow a lawn that measures 1, 235 square feet in 2.5 hours. At that rate, how long would it take her to mow a lawn of 4, 940 square feet?

 A. 9.5 hours

 B. 12 hours

 C. 12.5 hours

 D. 10 hours

21. Which equation represents the phrase "the ratio of 5 and a number is 65"?

 A. $\frac{5}{x} = 65$

 B. $\frac{x}{5} = 65$

 C. $5x = 65$

 D. $x = \frac{65}{5}$

22. What is the solution of the following equation?

$$12x - 80 = 100$$

 A. $x = 10$

 B. $x = 12$

 C. $x = 15$

 D. $x = 9$

23. What is the solution of the following inequality?

$$2x + 5 \leq 45$$

 A. $x \leq 20$

 B. $x \geq 20$

 C. $x \leq 15$

 D. $x \geq 18$

24. If $s = 8t - 5$ is an equation, then s is

   A. An independent variable.

   B. A constant.

   C. A dependent variable.

   D. A letter.

25. The linear model of the aircraft's descent gives its height as a function of time. Which of the following is true?

   A. The initial height is 0.

   B. The rate of change of the model is positive.

   C. The rate of change of the model is negative.

   D. None of the above.

The perimeter of the following rectangle is 84.

2x

5x

26. What is x?

   A. 8

   B. 12

   C. 6

   D. 13

27. What is the area of the rectangle?

   A. 360

   B. 180

   C. 168

   D. 160

28. A box that is 15 inches wide, 25 inches high, and 6 inches thick is to be wrapped in gift paper. How many square inches of gift paper are needed?

   A. $2,250\ in^2$

   B. $1,230\ in^2$

   C. $1,385\ in^2$

   D. $1,840\ in^2$

29. A map scale indicates that 3/5 inch on the map corresponds with 8 real miles. How many miles apart are two cities that are 4.5 inches apart on the map?

   A. 32 miles

   B. 36 miles

   C. 42 miles

   D. 60 miles

30. Michael runs 12 miles north and 5 miles east. What is the shortest distance he must travel to return to his starting point?

   A. 17 miles

   B. $\sqrt{17}$ miles

   C. 13 miles

   D. $\sqrt{119}$ miles

The mean of the following data set is 18.5.

17, 8, 28, x, 15, 31

31. What is x?

   A. 12

   B. 13

   C. 16

   D. 21

32. What is the median of the data set?

   A. 28

   B. 15

   C. 16

   D. 33

33. What is the range of the data set?

   A. 31

   B. 24

   C. 17

   D. 23

34. Which of the following cannot be a probability of an event?

   A. 88.8%

   B. 1.05

   C. 7/8

   D. 1/32

35. What is the probability of rolling an even number on a standard dice?

   A. 0.5

   B. 0.05

   C. 0.33

   D. 0.67

36. The probability of an event is always a number that is

   A. Less than 0.

   B. Between 0 and 1.

   C. Between 1 and 2.

   D. More than 1.

Answer Key:

1.	B	19.	D
2.	A	20.	D
3.	C	21.	A
4.	A	22.	C
5.	D	23.	A
6.	A	24.	C
7.	C	25.	C
8.	C	26.	C
9.	D	27.	A
10.	B	28.	B
11.	A	29.	D
12.	D	30.	C
13.	C	31.	A
14.	C	32.	C
15.	D	33.	D
16.	A	34.	B
17.	C	35.	A
18.	B	36.	B

REFLECTION ON LEARNING

After completing Practice Test #1, reflect on your performance by answering the questions below. Discuss your responses with your instructor or a classmate.

1. What questions did you answer incorrectly? List the question numbers.

2. Review the list. What types of questions (operations, measurements, algebra, geometry, data analysis, statistics, graph, pie chart) did you answer incorrectly?

3. Review each question you've missed. Why do you think you answered the question incorrectly?

4. Based on the questions you missed, what math functions or concepts do you need to study and practice more? List them.

5. Review the question you got correctly. What strategies or methods did you use? What did you do well?

6. After reviewing all the questions, what questions do you have for your instructor?

You have 70 minutes to answer 36 questions.

1. Jeff watched a turtle crawl 5.20 feet in one hour. The next hour, the turtle crawled 9/2 feet. How far did the turtle crawl in total?

   A. 14.4 ft.   C. 10.5 ft.

   B. 12.6 ft.   D. 9.7 ft.

2. Brianna can buy nine pencils for $3.78. How much does one cost?

   A. $0.45   C. $0.39

   B. $0.42   D. $0.40

3. Mr. Cooper worked 8.5 hours on Monday and 11.25 hours on Tuesday. How much longer did he work on Tuesday?

   A. 3.75 hours   C. 2.75 hours

   B. 3.5 hours   D. 3.25 hours

4. Alice studied for 21 hours last week. This week, she studied 25% more. How long did she study this week?

   A. 26.55 hours   C. 24.15 hours

   B. 27.75 hours   D. 26 hours

> In a box, there are 25 red marbles,
> 34 blue marbles, and 21 green marbles.

5. How many marbles are there in the box?

   A. 69   C. 78

   B. 74   D. 80

6. What percentage of the marbles in the box is blue?

   A. 45%   C. 42.5%

   B. 31.3%   D. 36.8%

7. What percentage of the marbles in the box is green?

    A. 25%

    C. 32.4%.

    B. 26.25%

    D. 33.3%

8. 40 % of x is equal to 3.2. What is x?

    A. 12

    C. 14

    B. 7

    D. 8

9. A jet travels 520 miles in 2.5 hours. What is the rate of speed of the jet?

    A. 120 mph.

    C. 208 mph.

    B. 260 mph.

    D. 300 mph.

10. Simplify the following expression:

$$\frac{(10^2 \cdot 10^2)^2}{(10^2)^0}$$

    A. 10

    C. 10^6

    B. 10^8

    D. 10^{10}

11. What is m?

$$(5^5)^5 = 5^{m+1}$$

    A. 9

    C. 20

    B. 12

    D. 24

Look at the following gas station receipt:

```
Passmore Gas & Propane
FG62326873455
3685 Charles Street
Livonia, MT
81065


9/12/2018   576646188
11:54 AM

XXXXXXXXXXXX2323
visa
INVOICE 831332
AUTH 138864

PUMP#24
Regular                19.58G
PRICE/GAL              $2.98

FUEL TOTAL              .?

    -----------
    Total =            ?

CREDIT
==========================
```

12. What is the price of a gallon of gas?

    A. $3.15   C. $19.58

    B. $2.98   D. $6.57

13. What is the total amount?

    A. $56.66   C. $58.35

    B. $58.74   D. $60.12

14. What is the cost of 20 gallons of gas?

    A. $55.00   C. $59.60

    B. $56.50   D. $58.98

15. If the price of a gallon of gas increases by 5%, what is the new price of a gallon?

    A. $3.13   C. $4.47

    B. $3.98   D. $3.57

16. Derrick walks 5.70 miles in 1.5 hours. What is the unit rate in miles per hour?

    A. 3.80 miles per hour   C. 3.55 miles per hour

    B. 4.15 miles per hour   D. 4.80 miles per hour

17. What is n in the following proportion?

$$\frac{7}{n} = \frac{91}{104}$$

    A. 8   C. 15

    B. 13   D. 7

18. Andrew had $140 to spend on nine books. After buying them, he had $32. How much did each book cost?

    A. $15   C. $22

    B. $11   D. $12

19. Grover wrote the following algebraic expression: $4x + 0.45x$. What would be the phrase that represents the algebraic expression?

    A. Four plus the quotient of 0.45 and a number

    B. Four times a number plus 0.45

    C. The sum between four times a number and 45% of the number

    D. The difference between four times a number and 0.45

> In 2014, the wolf population in a park was 235. By 2019, the population had increased to 360. Suppose the population continues to change linearly.

20. What is the rate of change of this linear model?

    A. 42 wolves per year   C. 35 wolves per year

    B. 20 wolves per year   D. 25 wolves per year

21. Which linear function represents the wolf population P, in terms of t, the years since 2014?

    A. $P(t) = 22t + 235$ C. $P(t) = 251t - 235$

    B. $P(t) = 25t + 235$ D. $P(t) = 235t + 25$

22. What is the wolf population in 2022?

    A. 435   C. 375

    B. 200   D. 422

23. When will the population reach 760 wolves?

    A. Year 2032   C. Year 2018

    B. Year 2035   D. Year 2026

24. If the area of a square is 121 square inches, what is the perimeter of the square?

    A. 22 in²   C. 42 in²

    B. 30.25 in²   D. 44 in²

The dimensions of a rectangular box are 7 in. x 13 in. x 5 in.

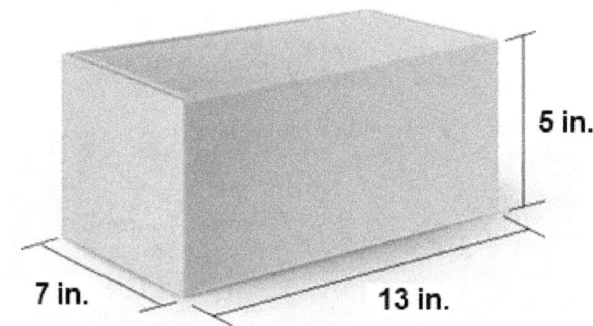

25. What is the total surface area of the box?

    A. 272 in²   C. 382 in²

    B. 375in²   D. 355 in²

26. What is the volume of the box?

    A. 455 in³   C. 462 in³

    B. 380 in³   D. 500 in³

27. A map scale indicates that N inches on the map corresponds with 5 real miles. Two towns that are 18 inches apart on the map represent 22.5 real miles. What is N?

    A. 6   C. 4.5

    B. 4   D. 5.5

28. What is x?

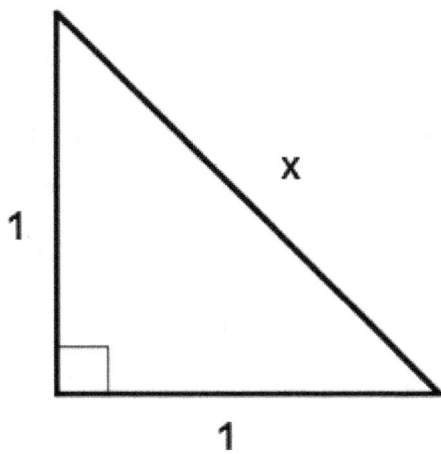

A. 2

C. $\sqrt{2}$

B. $\sqrt{3}$

D. 1

29. The leg of a right triangle is 12 inches and its hypotenuse is 13 inches. What is the measure of its other leg?

A. 11

C. 10

B. 9

D. 5

> The following data set represents the pulse rates
> (beats per minute) of eight students enrolled in a medical test:
> 62, 71, 87, 93, 70, 66, 74, 85

30. What is the mean pulse?

A. 81

C. 76

B. 72

D. 78

31. What is the median pulse?

A. 71

C. 77

B. 72.5

D. 73

32. What is the mode pulse?

A. 90

C. 84

B. 77

D. There is no mode.

33. What is the range of the data set?

  A. 31   C. 33

  B. 93   D. 79

34. What is the probability of rolling an even number on a standard dice?

  A. 25%   C. 75%

  B. 50%   D. 80%

35. Which of the following events has a probability that equals 1?

  A. A tail will come up if a coin is tossed.   C. The sun will rise tomorrow.

  B. A head will come up in a coin toss.   D. Lucy will eat a hot dog tomorrow.

36. There are 14 green cards and 26 blue cards in a bag. If one card is chosen at random, what is the probability of getting a green card?

  A. 0.65   C. 0.30

  B. 0.42   D. 0.35

Answer Key:

1.	D	19.	C
2.	B	20.	D
3.	C	21.	B
4.	A	22.	A
5.	D	23.	B
6.	C	24.	D
7.	B	25.	C
8.	D	26.	A
9.	C	27.	C
10.	B	28.	C
11.	D	29.	D
12.	B	30.	C
13.	C	31.	B
14.	C	32.	D
15.	A	33.	A
16.	A	34.	B
17.	A	35.	C
18.	D	36.	D

REFLECTION ON LEARNING

After completing Practice Test #2, reflect on your performance by answering the questions below. Discuss your responses with your instructor or a classmate.

1. What questions did you answer incorrectly? List the question numbers.

2. Review the list. What types of questions (operations, measurements, algebra, geometry, data analysis, statistics, graph, pie chart) did you answer incorrectly?

3. Review each question you've missed. Why do you think you answered the question incorrectly?

4. Based on the questions you missed, what math functions or concepts do you need to study and practice more? List them.

5. Review the question you got correctly. What strategies or methods did you use? What did you do well?

6. After reviewing all the questions, what questions do you have for your instructor?

You have 70 minutes to answer 36 questions.

1. An elephant weighs 5, 540 pounds. A horse weighs 870 pounds. How much more does the elephant weigh than the horse?

   A. 4, 760 pounds

   B. 5, 160 pounds

   C. 4, 670 pounds

   D. 6, 500 pounds

2. A family-size pizza is $32 and costs four times as much as a small pizza. Alex buys six family-size pizzas and five small pizzas. How much does he spend in all?

   A. $256

   B. $320

   C. $230

   D. $232

3. A factory produces 1, 565 bikes in a month. What is its annual production?

   A. 15, 520 bikes

   B. 9, 930 bikes

   C. 16, 260 bikes

   D. 18, 780 bikes

4. Dave wants to buy a laptop costing $870.95. He has $483.60 only in his account bank. How much more money does he need to purchase the laptop?

   A. $388.55

   B. $387.35

   C. $322.45

   D. $471.15

5. Janice is reading a math textbook. The first five chapters have 32 pages each. The last four chapters have 38 pages each. In total, the book contains 495 pages. How many pages are in the middle chapters?

   A. 183

   B. 216

   C. 180

   D. 191

6. Compute $3.45 \times (1 - 1/2)$

   A. 1.45

   B. 3.20

   C. 1.725

   D. 2.95

7. A pump moves 36 gallons of water in 3/4 hour. What is the unit rate in gallons per hour?

    A. 45 gallons per hour

    B. 48 gallons per hour

    C. 38 gallons per hour

    D. 42 gallons per hour

8. What is the value of Z in the following proportion?

$$\frac{12}{7} = \frac{Z}{91}$$

    A. 156

    B. 84

    C. 53

    D. 166

9. Maurice scored 55 out of 80 in a math test. What percent was that?

    A. 54.65%

    B. 65.50%

    C. 72.55%

    D. 68.75%

10. There are 25 students in a class. If 8% are absent on a particular day, what is the number of students present in the class?

    A. 2

    B. 14

    C. 17

    D. 23

11. What would the cost be for a $342 product selling at a 12% discount?

    A. $330

    B. $300.96

    C. $315.88

    D. $312.55

12. Simplify the following expression:

$$\left(\frac{8^8 \cdot 8^{18}}{8^{24}}\right)^{-1}$$

    A. 8

    B. 8^2

    C. 8^{-1}

    D. 1/64

13. What is p?

$$2^2 \cdot 2^p = 8$$

    A. 1

    B. 4

    C. 6

    D. 5

14. What is the value of *y*?

$$10\% \text{ of } y = 20$$

A. 30   C. 100

B. 80   D. 200

15. At a rental agency, renting a car costs a one-time fee of $34 plus $3 for every mile it is driven. Which of the following equations would be correct if it costs Bob $235 to rent a car?

A. 34x + 3 = 235   C. 235 – 3x = 34

B. 3x + 34 = 235   D. 34x + 235 = 3

16. What is the solution to the following equation?

$$6x + 60 = 600$$

A. x = 45   C. x = 90

B. x = 62   D. x = 120

17. Which of the following is a solution to the inequality?

$$8x + 5 > 45$$

A. x = 5   C. x = 1

B. x = 3   D. x = 6

18. Which of the following represents a linear equation?

A. $y = 1.7x$ C. $s = 1/t$

B. $x = y^2$ D. $y = x^3$

Look at the following triangle:

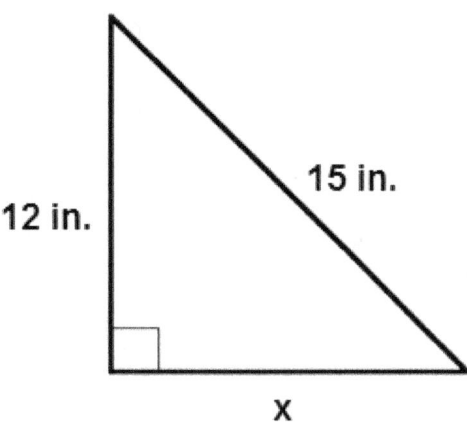

19. What is x?

    A. 27   C. 9

    B. 10   D. 5

20. What is the perimeter of the triangle?

    A. 52   C. 27

    B. 36   D. 34

21. What is the area of the triangle?

    A. 108   C. 62

    B. 90   D. 54

The following shape is formed by small cubes. The volume of each small cube is one cubic inch.

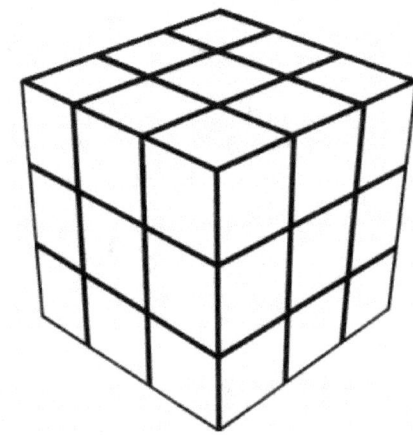

22. What is the volume of the shape?

    A. 27 in^3 C. 32 in^3

    B. 21 in^3 D. 21 in^3

23. What is the surface area of the shape?

    A. 81 in^2 C. 45 in^2

    B. 27 in^2 D. 54 in^2

24. A miniature model of a building is made using the scale 3 inches = 15 feet. If the height of the building is 40 feet, what is the height of the miniature model?

    A. 12 in.   C. 8 in.

    B. 10 in.   D. 6 in.

25. Which expression can we use to find the value of x?

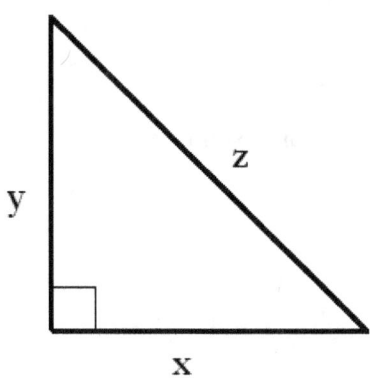

A. $\sqrt{y^2 + z^2}$

B. $\sqrt{y^2 - z^2}$

C. $\sqrt{z^2 - y^2}$

D. $y^2 + z^2$

The following data set shows the ages of nine students:

15, 18, 21, 17, 36, 15, 16, 19, 14

26. What is the mean of the data set?

A. 17

B. 19

C. 18

D. 21

27. What is the median of the data set?

A. 36

B. 15

C. 14

D. 17

28. What is the mode of the data set?

A. 18

B. 15

C. 19

D. 17

29. What is the range of the data set?

A. 22

B. 36

C. 25

D. 14

30. What is the outlier in the data set, if one exists?

A. 14

B. 36

C. 18

D. There is no outlier.

31. If the minimum value of the data set is removed, what is the median of the data set?

    A. 17

    C. 17.5

    B. 21

    D. 18

32. What is the probability of rolling a 2 on a standard dice?

    A. 1/2

    C. 1/6

    B. 1/4

    D. 1/5

33. What is the probability of choosing the letter "E" from the word GEOMETRY?

    A. 1/4

    C. 2/5

    B. 3/8

    D. 5/8

34. Which of the following describes the likelihood of this event:

    Rolling a number greater than 0 on a standard dice

    A. Unlike

    C. Certain

    B. Impossible

    D. Likely

35. Which of the following cannot be a probability of an event?

    A. 111%

    C. 1/433

    B. 0.007

    D. 1%

36. There are 12 pillows on a bed; 4 are white, 6 are blue and 2 are pink. What is the probability of picking a pink pillow?

    A. 1/12

    C. 1/2

    B. 1/3

    D. 1/6

Answer Key:

1.	C	19.	C
2.	D	20.	B
3.	D	21.	D
4.	B	22.	A
5.	A	23.	D
6.	C	24.	C
7.	B	25.	C
8.	A	26.	B
9.	D	27.	D
10.	D	28.	B
11.	B	29.	A
12.	D	30.	B
13.	A	31.	C
14.	D	32.	C
15.	B	33.	A
16.	C	34.	C
17.	D	35.	A
18.	A	36.	D

REFLECTION ON LEARNING

After completing Practice Test #3, reflect on your performance by answering the questions below. Discuss your responses with your instructor or a classmate.

1. What questions did you answer incorrectly? List the question numbers.

2. Review the list. What types of questions (operations, measurements, algebra, geometry, data analysis, statistics, graph, pie chart) did you answer incorrectly?

3. Review each question you've missed. Why do you think you answered the question incorrectly?

4. Based on the questions you missed, what math functions or concepts do you need to study and practice more? List them.

5. Review the question you got correctly. What strategies or methods did you use? What did you do well?

6. After reviewing all the questions, what questions do you have for your instructor?

You have 70 minutes to answer 36 questions.

1. Sean has driven 957 miles. He needs to drive 1, 348 more miles. How many miles will Sean drive in all?

   A. 2, 205 miles

   B. 1, 978 miles

   C. 2, 578 miles

   D. 2, 305 miles

2. There are 342 flowers, 113 of them are roses and the rest are sunflowers. How many sunflowers are there?

   A. 199

   B. 229

   C. 200

   D. 218

3. Mr. Jackson had 122 pencils in his classroom. He buys new boxes of pencils that have 18 pencils in each box. Now, he has 338 pencils. How many new boxes did Mr. Jackson buy?

   A. 15

   B. 13

   C. 12

   D. 21

4. Amanda has 8 backpacks with 25 books in each backpack. Greg has 7 backpacks with 34 books in each backpack. How many more books does Greg have?

   A. 38

   B. 27

   C. 41

   D. 36

5. Julia bought a bike for $344.78 and a backpack for $85.99. What amount did she spend?

   A. $388.45

   B. $468.38

   C. $430.77

   D. $466.67

6. Larry paid $45 for 12 cans of beans. What is the unit rate in dollars per can?

   A. $3.75 per can

   B. $4.68 per can

   C. $3.65 per can

   D. $3.70 per can

7. A new car can go 500 miles on 8 gallons of gas. How far can the car take you with 1 gallon of gas?

   A. 63 miles

   B. 57 miles

   C. 64.5 miles

   D. 62.5 miles

8. Martha paid $60 for a jacket that had been discounted by 25%. What was the original price of the jacket?

   A. $75

   B. $85

   C. $80

   D. $90

9. What is 1% of 1?

   A. 10

   B. 0.01

   C. 0.1

   D. 1.1

> Jennifer receives a monthly salary of $950 plus a 10% commission on items she sells each month. Suppose Jennifer's sales were $72, 500.

10. What is the amount of commission?

    A. $7, 000

    B. $7, 250

    C. $9, 500

    D. $8, 000

11. What is the gross pay?

    A. $8, 000

    B. $7, 500

    C. $8, 200

    D. $7, 950

12. Simplify the following expression:

$$\left(\frac{4^4 \cdot 4^8}{4^{10}} \right)^{-1}$$

    A. 1/8

    B. 4^3

    C. 1/16

    D. 16

13. What is x?

$$(8^8)^8 = 8^{x+8}$$

    A. 64

    B. 62

    C. 48

    D. 56

Look at the following receipt:

Boutique

4143 STATE STREET
LOS ANGELES, CA

REG#256   TRN#4616   CSHR#967110   Street#81

Helped By :   MARK

1	BLK ROCK	$49.99
3	TANK TOP @9.99EA	$89.91
1	SUMMER BLU	$16.49
1	T-SHIRT	$24.99
1	T-SHIRT SALE	$2.49

7 Item
SUBTOTAL: $183.87
TAX: $22.98
TOTAL: ?
CHARGE:

14. How many items were purchased?

A. 5   C. 8

B. 7   D. 4

15. What is the total amount?

A. $206.85   C. $216.85

B. $183.87   D. $198.76

16. What is the most expensive item?

A. Summer Blu   C. Black Rock

B. Tank Top   D. T-Shirt

17. What is the cost of one Tank Top?

A. $29.97   C. $29.87

B. $89.91   D. $28.99

18. What is the cost of six Summer Blu?

A. $89.98   C. $99.77

B. $98.94   D. $98.59

19. What is p in the following proportion?

$$\frac{4}{5} = \frac{52}{p}$$

A. 48   C. 65

B. 55   D. 62

20. Kayla wrote the following algebraic expression: $0.73x$. What would be the phrase that represents the algebraic expression?

A. 0.73 plus an unknown number   C. 73% of a number

B. 73 times an unknown number   D. 7.3% of a number

> A company introduced a new item. It costs
> $4,650 to develop the item and $32 to manufacture each item.

21. What is the linear equation that represents the total cost C(x) to produce x of these items?

A. C(x) = 32x + 4,650   C. C(x) = 4,650x + 32

B. C(x) = 32x - 4,650   D. C(x) = 4,650x - 32

22. What is the rate of change of this linear model?

A. $4,650   C. 32 items

B. $32 per item   D. $4,650 per item

23. What is the cost to produce 575 items?

A. $18, 400   C. $19, 990

B. $22, 505   D. $23, 050

24. What is the initial cost?

A. $4, 650   C. $4, 682

B. $32   D. $3, 232

25. If the perimeter of a square is 68 inches, what is the area of the square?

A. 49 in^2 C. 289 in^2

B. 144.5 in^2 D. 343 in^2

Look at the following figure. Each block is one square inch.

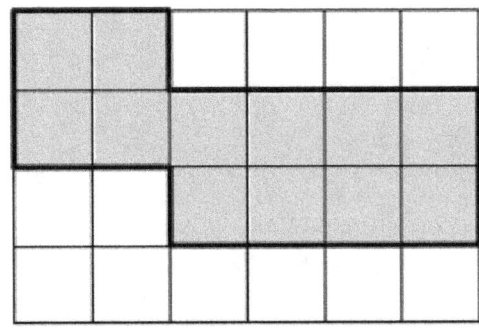

26. What is the perimeter of the shaded figure?

    A. 18 in.   C. 22 in.

    B. 15 in.   D. 19 in.

27. What is the area of the shaded figure?

    A. 15 in^2 C. 18 in^2

    B. 16 in^2 D. 20 in^2

28. A map scale indicates that 2 inches on the map corresponds with 5 real miles. Two towns that are *d* inches apart on the map represent 45 real miles. What is *d*?

    A. 14   C. 23

    B. 18   D. 31

29. What is *x*?

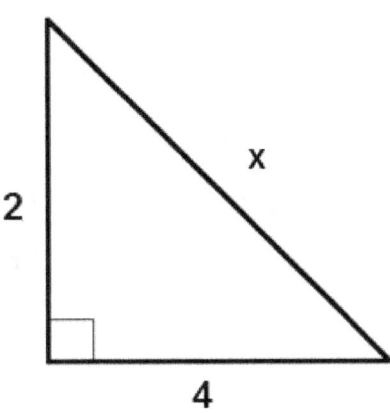

    A. 6   C. $\sqrt{20}$

    B. $\sqrt{6}$ D. 8

30. The leg of a right triangle is 3 inches and its hypotenuse is 5 inches. What is the measure of its other leg?

   A. 2 in.

   B. 3 in.

   C. 6 in.

   D. 4 in.

> The following data set represents the pulse rates
> (beats per minute) of eight students enrolled in a medical test:
> 73, 68, 72, 66, 72, 75, 74, 99

31. What is the mean pulse?

   A. 74.88

   B. 72.56

   C. 74.55

   D. 78.77

32. What is the median pulse?

   A. 71.5

   B. 72.5

   C. 74

   D. 73.5

33. What is the mode pulse?

   A. 73

   B. 99

   C. 72

   D. There is no mode.

34. What is the range of the data set?

   A. 66

   B. 99

   C. 33

   D. 72

35. What is the outlier in the data set, if one exists?

   A. 99

   B. 36

   C. 18

   D. There is no outlier.

36. If the minimum value of the data set is removed, what is the median of the data set?

   A. 72

   B. 68

   C. 75

   D. 73

Answer Key:

1. D	19. C
2. B	20. C
3. C	21. A
4. A	22. B
5. C	23. D
6. A	24. A
7. D	25. C
8. C	26. A
9. B	27. C
10. B	28. B
11. C	29. C
12. C	30. D
13. D	31. A
14. B	32. B
15. A	33. C
16. C	34. C
17. A	35. A
18. B	36. D

REFLECTION ON LEARNING

After completing Practice Test #4, reflect on your performance by answering the questions below. Discuss your responses with your instructor or a classmate.

1. What questions did you answer incorrectly? List the question numbers.

2. Review the list. What types of questions (operations, measurements, algebra, geometry, data analysis, statistics, graph, pie chart) did you answer incorrectly?

3. Review each question you've missed. Why do you think you answered the question incorrectly?

4. Based on the questions you missed, what math functions or concepts do you need to study and practice more? List them.

5. Review the question you got correctly. What strategies or methods did you use? What did you do well?

6. After reviewing all the questions, what questions do you have for your instructor?

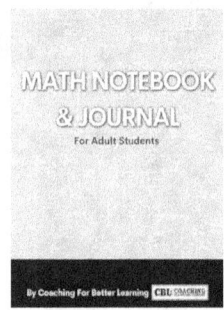

ABOUT CBL

CBL equips programs and instructors to increase student retention, learning—and success.

We do it by offering evidence-based systematic solutions, learner-centered teaching materials, instructor-centered training, and future-oriented strategies in adult education, workforce development, and vocational training.

We teach proven insights, knowledge, and skills that are useful to practitioners (instructors, administrators, and support staff).

CBL also takes pride in publishing student-centered textbooks designed to prepare learners for CASAS, TABE 11&12, HiSET, and GED assessments and assist instructors in covering course curricula and standards with confidence.

Our publications also include teaching guides, test prep tools, and study guides that foster reflective learning, ensuring sustained engagement in active learning. Find our meticulously crafted textbooks on our book page (cbledu.com) or major platforms like Amazon, Barnes & Noble, and Ingram Spark.

CBL also guides adult education and workforce programs in establishing robust professional development programs—training, peer-mentoring, coaching, community of practices (CoPs), and instructional systems— fostering a culture of continuous improvement and contributing to higher learner retention and success rates. We also offer workshops and PD sessions for adult educators and classroom instructors.

If you have suggestions or questions about instructional systems, textbooks, or student learning and retention, contact us today at teamcbl@cbledu.com or 410-960-4082.